Our Planet Earth

Deserts

by Sara Green

Blastoff! Readers 3

BELLWETHER MEDIA • MINNEAPOLIS, MN

Blastoff! Readers are carefully developed by literacy experts to build reading stamina and move students toward fluency by combining standards-based content with developmentally appropriate text.

Level 1 provides the most support through repetition of high-frequency words, light text, predictable sentence patterns, and strong visual support.

Level 2 offers early readers a bit more challenge through varied sentences, increased text load, and text-supportive special features.

Level 3 advances early-fluent readers toward fluency through increased text load, less reliance on photos, advancing concepts, longer sentences, and more complex special features.

★ **Blastoff! Universe**

Reading Level

Blastoff! Beginners — Grade K

Blastoff! Readers — Grades 1–3

Blastoff! Discovery — Grade 4

This edition first published in 2022 by Bellwether Media, Inc.

Library of Congress Cataloging-in-Publication Data

Names: Green, Sara, 1964- author.
Title: Deserts / Sara Green.
Description: Minneapolis, MN : Bellwether Media, 2022. | Series: Blastoff! readers. Our planet Earth | Includes bibliographical references and index. | Audience: Ages 5-8 | Audience: Grades 2-3 |
Summary: "Simple text and full-color photography introduce beginning readers to deserts. Developed by literacy experts for students in kindergarten through third grade"-- Provided by publisher.
Identifiers: LCCN 2021011385 (print) | LCCN 2021011386 (ebook) | ISBN 9781644875216 (library binding) | ISBN 9781648344893 (paperback) | ISBN 9781648344299 (ebook)
Subjects: LCSH: Deserts--Juvenile literature.
Classification: LCC GB612 .G743 2022 (print) | LCC GB612 (ebook) | DDC 551.41/5--dc23
LC record available at https://lccn.loc.gov/2021011385
LC ebook record available at https://lccn.loc.gov/2021011386

Editor: Rebecca Sabelko   Designer: Jeffrey Kollock

Printed in the United States of America, North Mankato, MN.

Table of Contents

What Are Deserts?

Deserts are large areas of dry land found around the world. They get no more than 10 inches (25 centimeters) of rain each year.

Deserts can be hot and sandy or cold and icy. Some are covered in salt.

Many deserts are **subtropical**. Daytime **temperatures** can soar to over 130 degrees Fahrenheit (54 degrees Celsius)! Nights are much colder.

Sahara Desert

Famous For

- World's biggest hot desert
- World's hottest desert

Type

Temperature

- Highest recorded: 136 degrees Fahrenheit (58 degrees Celsius) in El Azizia, Libya

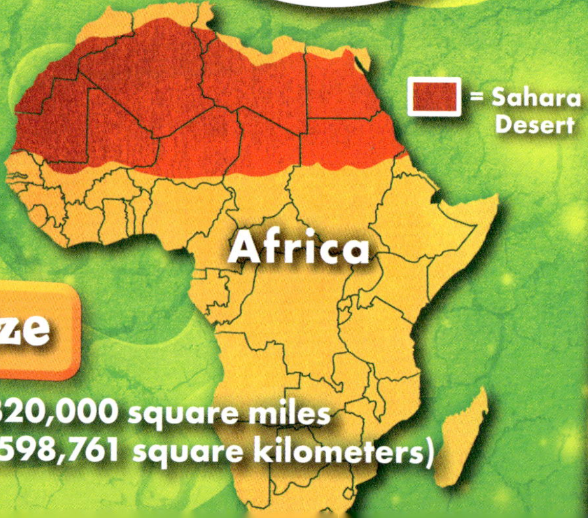

Rainfall

- Average around 1 to 4 inches (2.5 to 10 centimeters) each year

Size

- 3,320,000 square miles (8,598,761 square kilometers)

polar desert

In polar deserts, water is trapped in ice. These deserts are always cold.

Coastal deserts can form when ocean air becomes fog. Water droplets in the fog are too small to become rain. The land stays dry.

These deserts have warm summers and cold winters.

Atacama Desert

Famous For

- One of the driest places on Earth

Temperature

- Between 32 and 77 degrees Fahrenheit (0 and 25 degrees Celsius) during the day

Rainfall

- Averages less than 0.2 inches (5 millimeters) per year

Size

- Around 41,000 square miles (106,190 square kilometers)

Type

coastal

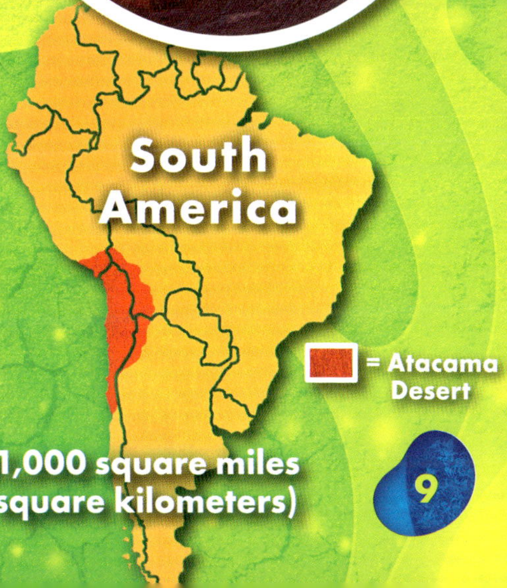

Many deserts are rocky and dotted with **scrub**. Deserts can also be sandy. Wind may blow sand into hills called dunes.

How an Oasis Forms

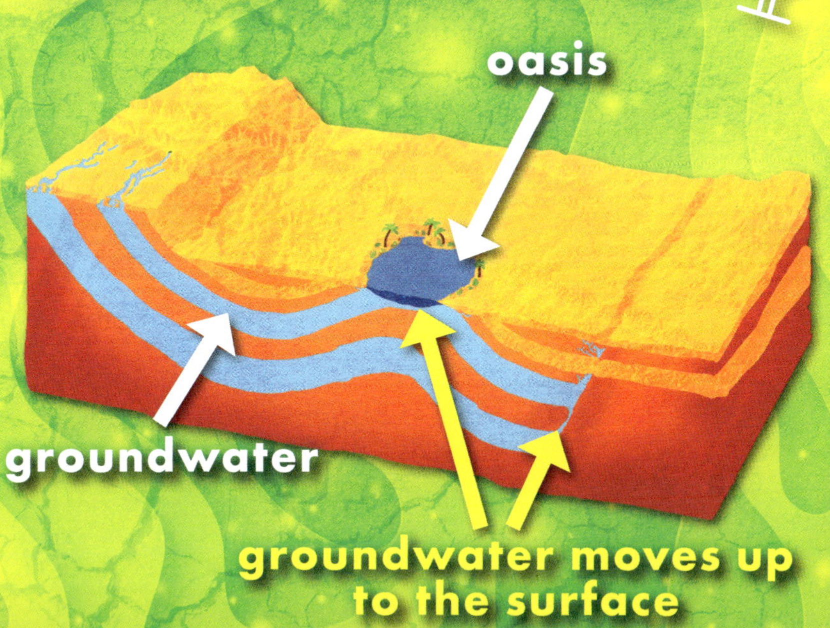

Sometimes **oases** are found where water is present. But they are often made by people.

Plants and Animals

Desert animals must survive with little water. Many get water from the food they eat. Some **rodents** even get water from seeds!

Fennec foxes fill up on juicy **insects** and mice.

Desert Animals

Desert plants must also live without much water. Many are **succulents**.

Succulents store water in their stems and leaves. Their roots soak up any rain that falls. Cacti have spines to keep thirsty animals away!

People and Deserts

More than one billion people call deserts home. Many live in cities. Some are **nomads**.

Deserts have many **natural resources**. People drill for oil and collect **solar energy**. They also enjoy camping and hiking in deserts.

nomads

solar panels
solar energy
farm

Human activity damages deserts. Cities **pollute** the desert air. **Climate change** causes water sources to dry up.

Farming and mining often destroy desert **habitats**. Off-road vehicles harm soil and kill plants.

How People Affect Deserts

- Desert cities pollute air and dry up water

- Climate change dries up water sources

- Off-road vehicles harm soil

People can protect deserts. Hikers and off-road vehicles can stay on trails made for them. People can build away from habitats that are easily damaged.

Deserts are important landforms. People must work to keep them healthy!

Glossary

climate change—a human-caused change in Earth's weather due to warming temperatures

habitats—lands with certain types of plants, animals, and weather

insects—small animals with six legs and hard outer bodies; an insect's body is divided into three parts.

natural resources—materials in the earth that are taken out and used to make products or fuel

nomads—people who have no fixed home but wander from place to place

oases—green areas in deserts that can support farming

pollute—to make land or air dirty

rodents—small animals that gnaw on their food

scrub—short bushes and trees

solar energy—usable power that comes from the sun

subtropical—relating to areas bordering the tropics; the tropics is a hot, rainy region near the equator.

succulents—plants that store water in their leaves, stems, or roots

temperatures—measurements of heat and cold

To Learn More

AT THE LIBRARY

Anthony, William. *Desert Food Webs.* New York, N.Y.: Kidhaven Publishing, 2021.

Castro, Rachel. *Amazing Deserts Around The World.* North Mankato, Minn.: Capstone Press, 2019.

Perish, Patrick. *Desert Tortoises.* Minneapolis, Minn.: Bellwether Media, 2021.

ON THE WEB

FACTSURFER

Factsurfer.com gives you a safe, fun way to find more information.

1. Go to www.factsurfer.com.
2. Enter "deserts" into the search box and click 🔍.
3. Select your book cover to see a list of related content.

Index

The images in this book are reproduced through the courtesy of: Anton Foltin, front cover; Marti Bug Catcher, p. 3; Zaruba Ondrej, pp. 4-5; Justin Fotos und Videos, p. 5; apstockphoto, p. 6; Yegor Larin, pp. 6-7; Smelov, pp. 8-9; Radek Borovka, p. 8 (fog); Jiann, p. 9; William Ducklow, pp. 10-11; Florence-Joseph McGinn, p. 12; Milan Zygmun p. 13 (gila monster); Hagit Berkovich, p. 13 (fox); Dennis W Donohue, p. 13 (roadrunner); Ondrej Prosicky, p. 13 (adder); Alexey Bagmanyan, pp. 14-15; gashgeron, p. 15; Dmitry Rukhlenko, p. 16; abriendomundo, pp. 16-17, 17 (solar panels); Repina Valeriya, pp. 18-19; Victor Mascheck, p. 18; Sean Pavone, pp. 20-21; arka38, p. 23.